# GLACIER'S SECRETS

## BEYOND THE ROADS *and* ABOVE THE CLOUDS

*by* **GEORGE OSTROM**
*with the* **OVER-THE-HILL GANG**

AMERICAN & WORLD GEOGRAPHIC PUBLISHING

## DEDICATION

*Glacier's Secrets* is dedicated to my sons, Shannon and Clark. I am thankful we shared wilderness adventures from the time they were small boys. Shannon became a skilled mountaineer as a youth and climbed in America and Europe, was on an army ski team, and served with NATO troops above the Arctic Circle before becoming a National Park Service ranger. After Marine Corps service, Clark has been a dependable and independent wage earner and husband. He also climbed with me, photographed the wild creatures, and fly-fished many a blue ribbon trout stream.

Their father loves them, and deeply admires their selfless courage in dealing with myotonic muscular dystrophy, a difficult physical adversary.

*Left:* What have we here? Looks like some cinquefoil, campion, several kinds of lichen, and a bunch of red rocks. If you walk up the big gentle ridge south of Scenic Point toward Mount Henry, this is what the ground looks like for three miles. When you can take your eyes off the ground there's far-out scenery too.

*Page 1:* Bob Zavadil strikes his pose in the notch between Matahpi and Going-to-the-Sun mountains, with Sexton Glacier below. St. Mary Lake is beyond. We climbed Matahpi that day, and found a different way down the northeast ridge to Siyeh Pass. This hike is a sleeper, rarely done, but anyone who likes scenery can never forget it.

*Front cover:* The Over-the-Hill Gang is headed for Shangri-La Lake on *the* goat trail off Bullhead Point. Notice I've conned Jack Fletcher into carrying my pinkish-red rope.

*Back cover:* At last we're on top of the Little Matterhorn, early September 1994. Seen from Going-to-the-Sun Road just east of Avalanche Campground, this thing looks like the Swiss Matterhorn. It is not a major Glacier peak, but the horn is sheer and it is a long ways from nowhere. *(see also page 66)*

*Facing page:* In Blackfeet lore, the Creator was Napi. This is Napi Point. Not a tough climb but we got into deep snow down in the trees and had to build a fire to get our boots dry. In the woods below we detected a mysterious aroma so exotic, it put any perfume in stores to shame. If we ever find the source, the foo-foo market is in for a jolt.

Text and photography © 1997 G. George Ostrom

© 1997 American & World Geographic Publishing

Write for our catalog:

American & World Geographic Publishing, P.O. Box 5630, Helena, MT 59604.

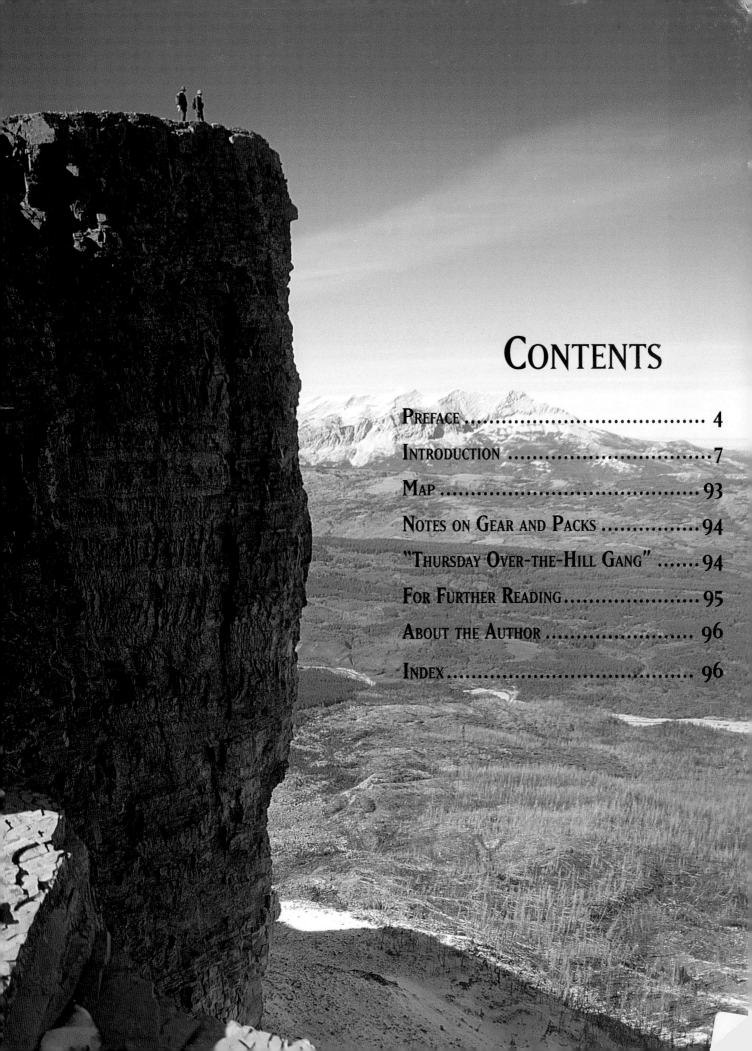

# CONTENTS

# PREFACE

This book is for everyone who loves wilderness, for those who cannot climb Glacier Park's great mountains and explore its million mysterious acres, for those who have been there, and for those who just need a start. Though very large, Glacier has only one road through it, Going-to-the-Sun, which is open half the year. There are a few short access roads on each side.

This leaves the vast majority of the park to those who hike. Much of that territory is viewable via 750 miles of trails. Beyond those "manways," there is no limit to exploring potential. Much of the un-trailed lowlands have difficult underbrush and downed logs to deal with, so moose trails are usually not happy trails. The best mountaineering is near and above timberline where the bighorns, grizzly bears, marmots, and mountain goats live. That is where most of these photos were taken.

The purpose here is a sharing of unique and diversified beauty, along with adventure in places seen by few. The scientific books on the park's biology and geology have been done and done very well. This is a humanized book about people visiting famous peaks or little known cirques and unnamed lakes, listening to waterfalls, and watching wild things live as they have for centuries.

Most of the photos were taken in the past fifteen years as the author hiked in excess of 4,000 mountain miles with the "Thursday Over-the-Hill Gang." The TOHG is an informal group averaging 350 hiker-miles each year and climbing 100,000 vertical feet in the process. Their median age is above sixty-five and there are two women regulars who are "much younger." They often have guests.

The author's approach to writing this book is one of good-humored narrative, conveying a sense of deeply enjoying what the mountains have to offer, but with a willingness to accept Mother Nature on her own terms.

CAUTION: With the fun, also comes potential trouble from weather, falls, avalanches, wild animals, and inadequate preparation. These threats can be lessened, and in most cases avoided, by selective reading, hiking with experienced people, and using common sense. Recommended is *A Climber's Guide to Glacier National Park* by Dr. Gordon Edwards. Besides its detailed information on climbing Glacier's peaks, the book is an excellent handbook on survival in what can at times be a dangerous world.

Becoming an "old timer climber," includes occasionally turning back, content in the knowledge that tomorrow the mountain will still be there. Other tips for avoiding trouble are found in this book's photo captions and text. Glacier Park maintains trained ranger teams that can and do make routine as well as spectacular alpine rescues. They are backed by both the ALERT copter-medic team from Kalispell and by Jim Krueger, a courageous and skilled pilot who keeps a contract 'copter at the park each summer. His feats are legend, but it is best to never need his services.

An absolute must for all true mountaineers is environmental awareness, a constant respect for the plants, animals, the earth, and your fellow man. The Thursday Over-the-Hill Gang supports the death penalty for anyone stepping on a moss campion.

There is a place filled with a jillion wild flowers, emerald ponds, red rock terraces, rich grass, waterfalls, and wildlife. Nobody but the TOHG goes there because it isn't on any trail, tucked away behind Reynolds Mountain and Heavy Runner. We call it "Eden." This isn't it. However, on the north side of Heavy Runner is a place of almost equal beauty and mystery we call "Eden North." This is it. We've never been down there but maybe someday. Stream is the Middle Fork of Reynolds Creek.

Ostrom strikes his climber pose on the Ptarmigan Wall, Elizabeth Lake to the north. Gordon Edwards gave me this fabulous hike for my fifty-eighth birthday on scary goat trail to Ahern Pass with the small original Gang in 1986. This shot taken on a rerun for my sixty-sixth in 1994. We met a charming gentleman on the trail below who was eighty-four. Asked him to join us, but said he had a hot date that afternoon back at the Many Glacier Hotel. Every man has his priorities.

# INTRODUCTION

Uncountable are the times I've stood beside a turquoise lake someplace in Glacier's solitude, gazed down from a peak, or sat quietly in a meadow filled with wildflowers and waterfalls and wished everyone on earth could share such magic. These are impossible daydreams, but I can share those wondrous places with words and photos.

In the late spring of 1996, I was asked to guide the National Symphony Orchestra members on a bus trip through Glacier and when that day was over, these world travelling musicians said they'd never seen such massive, overwhelming scenery in their lives. Yet some spoke of feeling like a kid in a candy store, allowed to "look, but don't touch." In response, the next day a mountaineering friend and I took a group where they could touch. It wasn't a long hike but it was an enchanting adventure serving up fresh lion tracks, curious deer, and a brief snow, dropping filmy curtains across towering cliffs at Avalanche Lake. Encores were discussed.

Later that summer, a back injury limited me to the trails for three weeks. A Thursday dawn found me on the Continental Divide, walking the Highline Trail to Granite Park where I met a doctor and his wife from Indiana. They'd planned to "just venture out a mile or so" from Logan Pass, but I invited them to make an eleven-mile circle with me to the Chalet and down to The Loop on Going-to-the-Sun Road. We had an unending show starring goats and bighorn rams above, dazzling flower fields on every side, and a golden grizzly with her two cubs feeding in a bright green meadow far below. We lunched where rock pikas piped and marmots whistled. Filled canteens from a crystal spring of ice water and watched an eagle ride the soaring thermals off Mount Gould.

All of this amid great purple peaks as far as the eye could see. That couple's gratitude for my sharing a day on the Garden Wall gave me a warmth and emotional high to cherish forever.

My blessing has been a boundless love for Glacier National Park since my parents first took me there in the summer of 1936. If I visit again tomorrow, I will feel the tingle of expectation. Thousands upon thousands of hiking miles, hundreds of climbs, and sixty-plus years of exploring have only made the wonder grow.

*Come along, share with me...*

⟶

Five mountaineers including my then-young son, Shannon, walk the north summit ridge on the top of Chief Mountain in the summer of 1972. I look at that climb like my military service, glad and proud I went, but not enthused about doing it again.

Chief Mountain appears absolutely unclimbable on the east side that everyone sees from U.S. Highway 89. It's an imposing 9,080–foot tower of rock, rising abruptly out of the wooded foothills; however a few experienced, and at least three inexperienced, people have climbed that face. Details about the mountain and climbing it are in the book, *A Climber's Guide to Glacier National Park.* The "Chief" stands slightly removed from the rest of the Lewis Range in the northeast corner of the park. In fact, the park boundary runs across the very tippy top. It is steeped in Indian lore and sacred to several tribes.

Besides the super, indescribably awesome, good views from up there, a couple of memories stand out about that climb up the west ridge. There were many in the group, Dr. Dave Downey with wife, Janet, and two kids, Ranger Riley McClelland with wife and kids, plus three cousins, and others.

After driving five miles, with fourteen people in my half-ton pickup, on a narrow, rutty, muddy, lousy road north of Otatso Creek, we attacked from the south. The two-mile walk to the base of Chief was fun, but the mile of steep scree slope was tough. From there the route led up through moderate cliffs of thoroughly fractured rock. Eventually we stopped to rest on a lofty ledge 200 feet below the summit. Janet Downey peeked into a crevice and exclaimed, "Hey! There's a golf ball." We retrieved a shiny new Titleist. After deep discussion, it was decided some real golf nut had carried the ball and a driver to the top of the mountain. There with great expectation he had teed up to make the longest drive known to western civilization and flubbed it. To keep from feeling bad the rest of the day, we also decided that anyone with that much imagination, surely brought an extra ball.

*This page:* The end of an August rain on the Highline Trail, one mile out from Logan Pass with six and a half left to the Chalet. Looking northwest through the clouds is seen the Glacier Wall leading from McDonald Creek up to Heavens Peak on the Livingston Range. Walking in clouds is usually great because it lends mystery. Occasionally there is too much mystery.

Never forget this trip. Ivan O'Neil and I talked our wives into an overnight at Granite Park Chalet after giving them absolute assurance that we would not see one single grizzly bear.

Ended up seeing eight, an all-time record for one hike.

The first seven were at a distance, mostly down in "Bear Valley" below the Chalet, but the eighth was "sort of in the trail" when we came out on Sunday. We were able to give him plenty of room by cutting across a switchback. Once at the car, our wives raised serious questions regarding their husbands' credibility. It took four years convincing them such a thing could absolutely never happen again.

*Facing page:* October on the Highline Trail to Granite Park, near the Grinnell Overlook trail junction. Unnamed peaks of the Continental Divide above.

Always a wonderful hike, the Highline Trail in summer leads to dozens of climbing destinations along the Garden Wall and offers great flower displays and varieties of wildlife. In September, it becomes almost deserted right when changing foliage supplies brilliant colors.

On a trip up from the Going-to-the-Sun Road in November of 1995, we found ice in the trail and on the return trip Gene Jacobsen fell down, hit the back of his head, and popped out both hearing aids in the snow. One guy said it was because all the air came whooshing out at once. A long search turned up only one aid, so the next spring when the snow had melted up that far, we hiked in a looked where Bob Zavadil has tied a little ribbon to mark the spot. Again no luck, so we went on up to the Chalet where the snow was still ten feet deep. On the return, Bob found the hearing aid. Gene put in a new battery and, in spite of seven months in the snow, it's been working ever since. Because of all the trouble he caused, there has been talk of making Gene wear little nets over his ears whenever we are walking on slick stuff.

*Above*: Blue grouse are usually seen just below timberline in the spring. The males establish courting areas, and spend a lot of time showing off. They strut around displaying colorful neck glands, eyebrows, and wing muscles, while fanning out their tail feathers. They also do some loud talking, while any females who might be interested sit around blinking their eyes. Sometimes people do this in bars.

*Right:* Gene Jacobsen's daughter, Karen, accompanied us on a Highline Trail hike in the summer of 1996. Good spring moisture had produced a bumper crop of wildflowers. We ate lunch right here among all the petunias and wild beasties, before moseying on into Granite Park Chalet and down to the Loop. Eleven miles of dirty work? Sure, but somebody…

Bighorn sheep scatter throughout the park above timberline in summer with the ewes and lambs going to different places than the rams. We now see more of these magnificent animals than we did thirty years ago. Bumped into this band while climbing on Pollock. Bighorns are not often near roads, because they hang high, but they do a lot of tourist watching. These were busy studying people way below on Going-to-the-Sun Road, until they got this chance to see one close up. Mountain in the background is Heavy Runner.

*Above:* Mountain goats are absolutely solid, hard muscle, and these kids are running around right from birth developing the condition and knowledge they need to survive. Saw these two very young ones near the outlet to Hidden Lake. No baby animals any cuter.

*Above (right):* The golden mantle squirrel is mostly an alpine creature, so we see them in high places. This one in Dawson Pass had just bullied three others away from something delicious it had found in a flower patch.

*Right:* Grinnell Glacier Trail in late afternoon. Notice notch just to the right of Mount Gould. There is a small glacier right on the Continental Divide named "Gem." We climb up the other side, from the Highline Trail. One time up there we started singing "Happy Birthday" to Elmer and Ivan, causing nearby goats to go berserk. They charged right through our group but nobody was rammed. That pretty much ended group singing when animals are around.

Here we have a considerable quantity of solidified water, and Jack Fletcher is explaining that next August this will come in handy for brushing teeth in Portland, Oregon, where he used to live. After snow closes the upper part of Going-to-the-Sun Road to vehicles, we like to hike up there. An early November shot just beyond the Big Bend, west side, the Garden Wall.

*Right:* Ivan O'Neil leads up final pitch to Mount Clements. Mountains to the east are Oberlin Ridge, Pollock, Siyeh, and Piegan. This is another jewel in the crown. Done it several times.

WARNING—The number one danger in our experience is falling rock, usually loosened by someone above. We've had minor injuries and a few very close calls which could have been fatal. Anyone causing a rock to fall must yell, "Rock!" Some places are bad enough we only allow one person at a time to move. Because of this problem, we limit the number of climbers, usually eight for precipitous routes.

*Below:* I know Ivan was standing in a herd of five ptarmigan when I clicked the shutter, but only four showed up in the photo. Maybe we could offer a prize to whoever finds the other one.

*Facing page:* Randy Heim and Jim Galvin on Glacier's most famous goat trail, north side of the Ptarmigan Wall. Lake Helen lies far below, the source of Belly River as it begins the long journey to Hudson Bay. This huge cirque receives never-ending water, falling thousands of feet from Ahern Glacier and smaller ice fields on Ipasha Peak, Mount Merritt, and Old Sun Glacier. If I were allowed to apply the word "awesome" to only one spot in Glacier Park, this is it.

The goat trail from Ptarmigan Tunnel to Ahern Pass was pioneered by Dr. Gordon Edwards. It begins in a saddle on top of the Ptarmigan Wall directly above the tunnel and offers four miles of breathtaking views of both the Belly River Valley and the Many Glacier country. It ends at Ahern Pass following a final few hundred yards across a monster wall with 3,000 feet of exposure. Not a good place for beginners. For those who reject that final scary ledge, there is an escape notch over the top of the Wall, and back down to the Ptarmigan Trail. Total trip is only sixteen miles.

On this day, we got a late start from Many Glacier so we used the escape notch return. On the trail near sunset, Bob Dundas was leading at Red Rock Point when a beautiful silvertip grizzly with two small cubs stood up and challenged with serious "whoofs." Bob instantly went from being our leader to being last in line. After a couple more bluff charges, bouncing on stiff front legs, the grizzly led her cubs across the trail and let us pass. I got my standard "semi-blurred griz photo."

Later, someone asked Bob why he violated the cardinal rule, "Do not run," and he said, "Well! I never saw anything move that fast." This from a guy who flew jet fighters in Korea and Vietnam. Another wondered why he didn't get out his bear spray, and he said, "Because it was in my backpack." Since then, Bob's pepper cannister is strapped to a front shoulder strap, and he claims to be the "fastest spray" west of the Mississippi.

*This page:* Examples of the brighter red Indian paintbrush. Sometimes we'll see wide color variations of these cheerful plants all in one patch.

*Above:* Here we have very short plants not in my books, but they exist. See a lot of 'em. The almost constant wind and short growing season keeps most plants low in high country. Thousands of acres of evergreen trees never get two feet high. Examples are on the northwest side of Little Chief Mountain above St. Mary Lake. They just hang on, huddling close to the earth. Swiss, Austrians, and Germans call them *Krummholtz*.

*Above (left):* A pine marten with prey.

*Top of page:* Got driven off a mountain by storms this early June day, but salvaged the afternoon by hiking along Lake McDonald where this young muley buck peeked at us from the bushes. Also found half a dozen delicious morel mushrooms. It is legal for people to eat park mushrooms and berries, but not flowers and shrubs.

Taking a break at Ptarmigan Lake, four miles from the trailhead. The Tunnel is located high up on the distant rock wall. It comes out the other side in the middle of a sheer cliff, so the trail there had to be blasted from solid rock. A spectacular bit of engineering, and that one first view down Belly River across Elizabeth Lake and Natoas Peak towards Waterton Lakes National Park in Canada pays for the trip. Take a short walk north from the tunnel and you can look westward to Helen Lake, Mount Merritt, Old Sun Glacier, and Ipasha Peak, without doing that scary goat trail.

*Above:* Yellow on green. Flowers and lichens.

*Facing page:* Goat Lake, nestled in one the most stunning "unseen" basins in the park. View is looking east to Mount Otokomi and on to the Great Plains toward Chicago. There are two major summits to Goat Mountain and this shot was taken from between them. Reaching the top requires no technical climbing if approached from the west up *Sound of Music*–type alpine slopes above Baring Creek.

The entire trip to Goat Lake is a joy, with every kind of natural miracle Glacier has to offer, flower displays, wildlife, water falls, glaciers, lichens, and all against the immediate backdrop of big peaks, Going-to-the-Sun and Matahpi, plus all the mountains along the south shore of St. Mary Lake.

Returning from one trip up there, we were resting above Baring Creek, soaking up the beauty all around, when an Australian girl with us suddenly leaped in great excitement yelling, "There goes a dingo! Ayev spent me life ohping to see a Dingo, and ayev ad to vist Montana ta find one."

We had that day observed bighorn sheep, goats, and deer, but our guest's excitement clearly showed there can still be great wonder in a common coyote.

Few people have visited Goat Lake itself, but some of the gang pioneered a route into it in 1995 by leaving Going-to-the-Sun Road near Dead Horse Point, ascending the southeast buttress of Goat Mountain to the approximate altitude of the lake and then carefully circling through the moderate cliffs above Rose Creek. The gang's triumphant entry to the emerald lake amphitheater surrounded by 2,500-foot walls beneath gigantic peaks, produced hushed, profound words like, "Oh golly," "Holy cow," and "Wow."

Walking east on the high colorful ridge between Chief Lodgepole and Painted Teepee peaks, daughter Heidi Duncan called this "A Sky Path Through Paradise." Off to the distant left are Mount Sinopah and the massive 9,600-foot Rising Wolf Mountain, north of Two Medicine Lake. Distant right shows Never Laughs, Appistoki and Mount Henry. Not visible to the immediate left is a 1,200-foot cliff ending in Cobalt Lake. Easiest way up here to the Continental Divide is by trail through Two Medicine Pass, then hang a left.

In the Two Medicine and Cutbank east side valleys there are many of these "peak-bagger bargains." If you climb one mountain, you get another one free. The entire Two Medicine area is a favorite. There are so many varieties of peaks, passes, lakes, and "bargains." Besides that, one big mountain named Spot has several summits and the south ridges get clear of snow early, enabling first climbs of the spring. Lot of wildlife here. In summer the boat runs up the lake and gives an early morning boost into the backcountry. Conversely it can save a couple of miles at the end of long days, if you get back in time.

There is a super "crossover" trail hike between Two Med and Cutbank Ranger Station via Pitamakan Pass. Groups start from opposite ends and trade car keys in the middle. You have to either do this with people you trust, or who have a nicer car than you do. Another dandy hike on trail all the way is the Pitamakan Pass, Cutbank Pass, Dawson Pass Loop. It can be done in either direction but I prefer hitting Dawson last because you can catch the boat back; and with proper planning have "something cold" waiting for that relaxing water trip back to your car.

Pikas are high mountain farmers. They devote brief summers to gathering hay which they pile on rocks to cure before storing for winter. If storm clouds appear they rush to put the hay away, then when the sun shines, bring it back out. They don't really hibernate, just spend nine months in their dens. Don't try this at your house. According to a lady we took to a pika colony, "They are just absolutely darling." Dale Haarr, who comes from Two Dot, Montana, can talk to them.

These bighorns are pals in the posies now, but in ninety days they'll be battling each other for love on the ledges, and there aren't many rules. Watching a ram go soaring off a cliff in twenty- to forty-foot drops, seeming to float while barely touching here and there is unbelievable. I've seen them drop 300 feet in seconds, come to a well cushioned stop, then calmly start grazing. Must have steel springs for legs.

*Above:* Certain lichens grow only half an inch in a thousand years. You may be looking at a living thing dating back 10,000 years to the ice age. My ranger son says he believes this organic gathering is a good example of plant symbiosis in Glacier. I just know it is one of the most beautiful little compositions of nature I've ever seen. The wondrous creation is on a cliff at the 8,500-foot level, south side of Reynolds Mountain. Taken August 19, 1993.

*Facing page:* Ted Rugland is not dawdling, he just likes to see everything well before moving on. Almost-a-Dog and Citadel mountains rise beyond St. Mary Lake as Ted moves upward across *The Sound of Music* slopes above Baring Creek, inexorably toward the summit of Goat Mountain. View is looking southwest in a drainage where all waters flow to Hudson Bay. Little kids stop Ted in stores and tell him what they want for Christmas. Baring Creek has its beginnings in the north at Siyeh Pass, and we recall several preventable accidents which happened to hikers in that area. Remember three people being hauled out in one day after suffering injuries from playfully sliding down snowfields. One hit rocks at the bottom, another went over a small cliff, and a third cut his arm badly on a sharp piece of rock imbedded in the ice. Sliding on snowfields is fun if you know what you are doing, but the danger is not apparent to the uninitiated.

Bob Zafadil on the ridge between Chief Lodgepole Peak and Grizzly Mountain. The thing to notice is not Bob's old hat, but rather the brilliant glow of red and yellow wildflowers, growing in the rocks. The miracle is how all those dainty, delicate blooms manage to live in that harsh climate with only sixty frost-free days a year. They are a constant summer delight, from the highest peaks to the deepest canyons.

*Above:* Saw a young mule deer buck in the bushes and was maneuvering for a picture when two big ones walked out. Had to get the film developed before discovering the buck on the right has his ears on backwards. As for the little buck? He managed to get in the photo anyway but you have to look twice.

*Right:* I was leading when this middle-sized black bear made a snort and stood up. I think it's a pretty good shot considering it was done left-handed, while I went for the bear spray. As a newsman, I have interviewed eleven people who successfully stopped a charging grizzly with Counter Assault pepper spray, but Ivan stopped six members of the Over-the-Hill Gang with one little "test squirt" into the wind.

Big six-point elk killed by another bull on the Belton Hills winter range in Glacier. After examining him, son Shannon and I figured he may have inflicted mortal wounds on his adversary. There was blood on most of his antler tines, far enough up to indicate deep puncture wounds inflicted on the other animal. We have found several of these mating season victims in the past. The natural world shows us many harsh realities, which are an integral part of life in the wilds. *(Photo by Shannon Ostrom)*

That old bighorn ram on the left has fought for many mates. Note the broomed-off horn and broken nose that makes him look like an old prizefighter. Big rams often come out of the late fall rutting season with injuries rendering them easier prey for predators. "King of the hill" may be where the action is, but it can make for a shortened life.

On the trail to Pitamakan Pass. Beargrass is the predominant flower in this shot. Some years are much better than others for these big white blooms; 1988 was a good one. Individual plants are on a seven-year cycle and big game animals eat 'em. Have mostly caught mountain goats doing that. People are not to pick any living thing in Glacier Park, but if a tourist had ever eaten one, there is no record of him or her being apprehended.

Glacier has scads of unnamed lakes and ponds, including some fair-sized ones. While son Shannon was working the West Entrance gate, one visiting lady asked if there were any "undiscovered lakes" she could look for.

We've named two high lakes in this drainage. One is "Over-the-Hill Lake." The other we call "Echo" because that's what it does. On the hike in there, one of the group stopped to take a picture of the Gang all strung out below on a ridge. That's when he noticed the fourth guy down was a friendly grizzly bear. Officially named lakes in this beautiful canyon include Old Man, Young Man, and Boy.

*Above:* Picture of Jack Fletcher taking a picture of a pair of ptarmigan: but if there are two birds in this photo, I need new glasses. One is easy to find. Returning to Logan Pass from visit to Eden and climb on Heavy Runner.

*Left:* Hanging Gardens looking northwest toward Mount Oberlin and the Garden Wall. Did six miles that early August day in 1996; up to the Reynolds Notch from Hidden Lake Overlook, around the cliff tops above "Eden North," then back to Logan Pass in the late afternoon. Ivan wanted to do that with Dr. Hi Gibson and the Thursday Over-the-Hill Gang founder, Ambrose Measure. It was an early present for Ambrose's eighty-eighth birthday in December. Tough getting in here during December, and anyway, Ambrose would be busy with downhill skiing at that time. Saw a mule deer buck eating the flowers here.

*Right:* Mountain goats are the number-one companions of climbers in Glacier. This one was taking his mid-morning siesta 8,000 feet up on Mount Reynolds. To his right is the south ridge of Bearhat Mountain with Mount Brown beyond. We both wondered what the other one was doing up there.

*Below:* From the summit of Mount Cannon, all the peaks surrounding Logan Pass can be seen. We are looking at the western spine of Clements with Going-to-the-Sun beyond. In the center are Heavy Runner and Reynolds with Hidden Lake below.

On this August day in 1988, one of the originals of the Over-the-Hill Gang, Harry Isch, wasn't feeling well when we reached the saddle between Cannon and Clements so Dr. Hi Gibson and Pat Gyrion started down with him. The rest of us continued climbing. About half a mile west of Hidden Lake Overlook, Harry was joking about somebody else having to carry his pack when he fell and died instantly of a heart attack. We all felt a tremendous loss, but agreed there couldn't be a better way to go. "A peaceful death, among good friends, in a beautiful place."

Hank Good on Mount Cannon the day Harry Isch died. "Colonel" Hank was one of those rare pilots who flew many combat missions in World War II, Korea, and Vietnam. He was nearing seventy when he began climbing with us in Glacier, and made it up a lot of big hills when he had physical problems. Just before he died, he told me they were some of the best days of his life.

Some say this is a high country whitebark pine, killed and cured by the howling winds. When I was a boy, a ranger told me they were called silent dog trees because "they have no bark." Dean Dahlgren and Bob Zavadil stopped to rest with this dandy specimen beside Scenic Point Trail, just above Two Medicine Ranger Station. Mount Appistoki is the background. We had just climbed Mount Henry where I fell near the summit, causing a compound fracture of my left candy bar.

Glacier Park is not all hang-by-yer-nails. There are many summers' worth of beautiful trails of varying length and steepness, offering fun and beauty beyond measure. A good example is the Scenic Point Trail, which takes you up through some bighorn sheep country, past silent dog trees, topping out on a timberline ridge where, if they just had a hill over fifty feet high, you could probably see North Dakota.

Little kids hike here all the time, not to mention some old men. That relatively short hike will also get you in a vantage position to look back west up the Two Medicine Valley with Glacier scenery you can hardly believe.

Any off-trail exploring should not be done alone, without knowledge of what is involved, and, of course, never in a manner that damages fragile environment.

Female grizzly with a big cub has just given up on attempts to kill a bighorn ram for supper. She's saying, "To heck with it," but the kid is saying, "Ah Ma, couldn't we give it one more try?"

My park ranger son Shannon, son-in-law Scott Duncan, friend Mike Schlegel, and I were photographing a big herd of rams on Mount Henkel in early spring when these grizzlies circled downwind and tried to pull off two different stalks on the sheep. It went on for an hour while we watched from a cliff ledge. The sheep seemed to assign four or five of the biggest rams to keep constant track of the griz while the rest went on with life as usual. After giving up the hunt, the bears were hot so they went to a snowbank and pressed down into it to cool off.

*Facing page:* Iceberg Lake is visited by many day-hikers coming up on the maintained trail from Many Glacier, but only a few get this unique angle. Jack Fletcher is coming down *the* mountain goat trail from Shangri-La.

Shangri-La is a large, treed meadow with a small lake in the middle, hanging loftily on the side of Bullhead Point, which is a buttress coming off the mightiest peak in the Swiftcurrent Valley, Mount Wilbur. There are three ways to get to Shangri-La but by far the most spectacular is to act like you are making a "direct challenge" up the east face of Wilbur via the waterfalls northwest of Red Rock Lake and then "chicken out" on the top of Bullhead Point. Then you must find the only goat trail down through the cliffs around the southeast end of Bullhead and—*Shazam!*—there you are in Shangri-La. It is just seven miles back to your car, providing you find *the* goat trail to Iceberg. Someplace up there Jim Galvin's gold and jewel-encrusted retirement watch got lost. He thought about going back to work for the company to get another one, but realized he'd be ninety before he earned it.

Years ago in the early spring, I made my first trip to Shangri-La with the gang but only Ivan O'Neil and my son Shannon got up there with me via the ridge and cliffs north of Fisher Cap Lake. The descent to Iceberg Lake was in deep snow. Then, adding to our wilderness experience, the footbridge over Wilbur Creek was gone, so we spent half an hour finding rocks to build an island. It was eight feet across so we built the island out there four feet. When it was done, I went first and jumped with such great enthusiasm I sailed over the rock pile and went to my waist in ice water. I've had many adventures up there in the fabulous Many Glacier country, but that was by far the most refreshing.

*This page:* Life is not easy for young mountain goats. They have a high mortality rate in their first year. Falls from cliffs are the chief danger. Watching them playfully chasing each other, dashing hither and yon where a slip could be fatal, makes you wonder how even one survives the summer. We see the kids taking naps on high ledges, and the nanny always stays between her baby and the cliff edge.

*Above:* Do not love daughter Heidi more than daughter Wendy, but Heidi turns up for Thursday hikes. Maybe it's because when I am up in the park, Wendy has to be back in Kalispell

anchoring Montana's top radio news department. Anyway, on this day, the Over-the-Hill Gang split up and climbed four major peaks of the Crown. Our group did Mount Oberlin where Heidi posed in this field of asters. Garden Wall with Highline Trail is in the background. Wendy and youngest son, Clark, climbed Oberlin with me many years ago. They were in high school, and we took some dandy "scare the heck out of Mom" photos.

*Left:* Cinq blossoms in campion.

Sitting on the knife-like top of Bullhead Point, two of the more mature members of the TOHG are watching three younger ones up on the northeast ridge of Mount Wilbur. This is what happens when we take along kids in their fifties, who think they can do anything. Conquering Mount Wilbur safely involves technical climbing and mucho savvy. The three adventurers in this case made it up to that flat spot with the vertical rock tower. While they were coming back down to join us, one guy's hat blew away forever.

*Above:* Can a rock ptarmigan look like a rock? This one can. Met up with him on the Continental Divide above Two Medicine Pass. These alpine grouse are survivors, and they must owe it all to their changing colors throughout the year, because they seem dumber than the rocks they sit on. In winter they're snow white. This is in July.

*Right:* Dean Jellison sits atop Bearhat Mountain looking east across Hidden Lake to Mount Reynolds, and other peaks of the crown. Dr. Edwards feels the views from here are among the most interesting in the park: "Almost every major peak in the park is visible from here except for those concealed by massive Mount Jackson. In every direction there are jagged walls and pinnacles, splendid cliffs and densely wooded valleys." It was Mount Jackson that inspired George Bird Grinnell in 1891 to first call this area the "Crown of the Continent."

44

*Above:* Shannon Ostrom stands in "the keyhole" high on Reynolds Mountain. It was his first major climb and we went with Dave Downey and his son, Mark, in 1970. Used ropes to teach the boys about safety and they learned it well. Since this picture was taken, the top of the keyhole has fallen out, reminding us of the constant change, even in rock.

*Facing page:* Sometimes I think this is my favorite photo of Glacier Park. Taken in October from the Highline Trail north of Granite Park Chalet. Forest fire smoke caused haze. Autumn colors are out and view is south down the Continental Divide to Mount Gould. What a wonderful area and memories of a hundred adventures.

*Facing page:* A lot of gray hair on the southeast ridge of Cracker Peak. Boulder Lake is below an unnamed ridge. Part of Goat Mountain, and a bit of St. Mary Lake show through the notch. Divide Peak on the horizon marks the Hudson Bay Divide on the park's eastern boundary.

Hiked up here via Preston Park from Siyeh Bend because I had a plan to climb Cracker Peak and walk the high ridge to Wynn Mountain, ending up that evening at Many Glacier, many miles away. Saw wonderful things the first few miles, including a massive bighorn ram, million wild flowers, then a newborn lamb with its mother. As she was feeding below us, two goats came out of the cliffs. The ewe didn't notice the little guy go see what those big white things were. One of them threatened him and he took off for mamma. We think she bawled him out.

The "really good plan" ended when we hit a short but dangerous ledge. While I was trying to figure out a rope belay, two lightning strikes came down on the 10,000-foot summit of nearby Mount Siyeh, and we took off for lower country. The storm passed over and we caught up with a California couple who had been watching Boulder Lake with binoculars and counted fifty-four elk down there. When we got back to Siyeh Bend our cars weren't there of course, because we'd sent them over to Many Glacier with three members who hadn't cared much for my plan in the first place. Got home very late, but wife, Iris, said I had to take a shower anyway.

*Below:* Here we have the typical bighorn sheep day-care center. Four kids were behaving themselves fairly well, but two others kept wandering off to sniff balsam root flowers. Besides not getting in the class photo, they probably missed their naps and were fussy all afternoon. Ewes without kids of their own often volunteer for day-care duty.

*Above:* There is no color known to man that is not reproduced from its most vivid hues to the most delicate shades and tints by the plants and rocks of Glacier Park. Have no idea what kind of petunias or shrubs these are but there are thousands of this and other variations growing in the rocks on ridges. This is a July model on the west flank of Bear Head Mountain.

*Right:* Walter Bahr and Ivan O'Neil believe this could be an exciting new scenic route to Otokomi Lake, and try it every few years along the unnamed ridge east of Siyeh Pass. Almost made it one time, so who knows? For those not overly thrilled by strolling narrow spines 8,000 feet in the air, there is a nice trail to Otokomi Lake up Rose Creek from Rising Sun campground. Walt, Ivan, and I, all fibbed about our age during WWII to obtain early employment with the local Forest Service. That's why we are above-average route finders.

*Above:* Ed Mall and Jim Galvin eagerly arriving at the remote Windmaker Lake. We'd heard there were nice trout in those cold deep waters. But just as I snapped this shot, my friends spotted three grizzlies above us. 'Twas a golden tan female with two cubs almost as big as she. After a while the three bears went up the backside of Mount Wilbur and we had lunch on the beach. That was interrupted by a young bull moose wading out to feed. When we were finally ready to start fishing, the griz came back—not real close, but not real far either. The outcome was no fish but a wonderful day.

*Facing page:* Best climb for me in 1996 was my annual July birthday outing. This one went 4,000 feet up Apikuni Mountain. On the way through the Natahki Basin I reminded the gang that the finest cold water spring in the world was right here. Heidi Duncan filled her canteen as Bob Zavadil and Ed Mall stood in wonder at the wildflower display.

The Blackfeet Indians used this high hanging valley as a "horse hospital," according to records of James Willard Schultz. They could put horses in there and block off escape with a simple pole barricade above Apikuni Falls.

Apikuni was Schultz's Blackfeet name. Working with George Bird Grinnell, he is recognized as co-founder of Glacier National Park. Other than that he was just your typical educated eastern kid who came out west to shoot a few buffalo, but married an Indian girl, joined her tribe, went on the warpath, and wrote about sixty books.

If you leave Logan Pass early, you can go to Hidden Lake Overlook, down 1,100 feet and around the lake, up a game trail through the eastern cliffs, thence along the giant ridge between Bearhat and Dragon's Tail, down to Floral Park, up the Sperry Glacier moraine, across the foot of the glacier to Comeau Pass, and down to Lake McDonald Hotel in one day. It is twenty-two miles with more down than up. This shot is Heidi Duncan and Elmer Searle entering Floral Park where Heidi asked me if giving blood the day before would affect her stamina.

The first time I tried this was with Hal and Jimmy Kanzler in 1964. Jimmy was in high school. We went directly across the snow-covered glacier with the boy out front on a rope. Hal said if Jimmy went through a snow bridge and into a crevasse, we could pull him out and find a safer route. He did and we did.

Jim's younger brother, Jerry, was one of five young men killed on Glacier's highest peak, Mount Cleveland, during a dangerous late-December climb in 1969. They were caught in an avalanche and carried 2,000 feet down the west face. Their bodies were not found until the following summer. I went up searching on June 26, accompanied by a ranger, and was turned back in late afternoon by bad weather and falling rock. The following Sunday, a four-man team of Waterton and Glacier rangers found the bodies frozen in deep ice on a ledge above the highest waterfall on the face. That was about 200 feet above where I turned around.

When Jerry's body had been removed from the ice on July 3 and helicoptered to a meadow near Chief Mountain, I put him in a plain pine coffin his mother had ordered, placed his favorite peace medal upon his chest, and brought him back to Kalispell for burial. One of the most difficult things I have ever had to do. The next week, with a friend, Bill Martin, I returned to Mount Cleveland and the scene of the boys' deaths, just thinking and trying to make some sense out of the tragedy.

Jimmy Kanzler has spent his life climbing and, with Ranger Terry Kennedy, made a first ascent of the north face of Siyeh in 1974. Jim compared its technical difficulty and danger to the infamous north face of Eiger Mountain in Switzerland.

That type of climbing is not what this book is about, but I do understand young men and mountains.

*Above:* I have approximately 8,236 pictures of Ivan O'Neil's backside heading up some trail, scree slope, or cliff. Thankfully this one caught a profile. The Hanging Gardens' 2,000 fabulous acres almost surround the Logan Pass Visitors Center, with the Continental Divide in background. Beginning hikers can see the area well by going up the boardwalk to Hidden Lake. Ivan saved me from drowning when we were juniors in high school, and there is still a scattering of people who can't forgive him.

*Right:* Mule deer are high-country animals in August, especially the bucks. This one in velvet antlers was having very colorful salad in the Hanging Gardens. Bright red flowers are Indian paintbrush. Earlier in the season, this area is bright yellow from the glacier lilies. Summer doesn't stay long but is very busy while it is here.

*Facing page:* South face of the Dragon's Trail. Four of us were envious when we looked over and saw three of the gang had found an easier way up cliffs that had us stranded. The Dragon's Tail is not on maps. I started using those descriptive words in the 1960s after viewing it from Mount Reynolds. Later, Dr. Edwards found a way to get on top and used the name in his *Climber's Guide.* The Glacier Mountaineers Club also uses Dragon Tail. YES! There are three people in this photo.

*Above:* A wondrous moment. Glacier is home for many whitetail deer on the west side. This doe carefully peered up and down the North Fork River before giving a safe signal. Then, flashing from the brush, came two young fawns and for ten minutes it was a non-stop whirlwind of spotted tan excitement—exuberance out of control. The little guys were in and out of the brush, leaping in the air, jumping in the water, wheeling to chase each other round and around. The brief and joyful celebration of life was over too soon. A mysterious order (from the doe?) and the fawns seemed to freeze in mid air, then shyly obedient, followed their mother back into the silent concealing forest.

*Left:* Chipmunks dash hither and yon so fast they are almost impossible to photograph, but I catch one now and then. There seem to be two slightly different kinds of these charming little hibernators.

Grinnell Point is a spectacular tower of rock directly across Swiftcurrent Lake from the Many Glacier Hotel. While sitting on the porch of that imposing log structure, the Point appears un-climbable except by experts with rope and hardware, but, *au contraire*, it is 2,630 feet above the hotel, and I have personally conducted "almost elderly" ladies and gentlemen to the top on several occasions. It is not a stroll, and people have died up here, but done carefully and slowly, it is one of the very best moderate climbs in Glacier. View here is across the Great Plains in the general direction of Minneapolis.

Climbing into the Iceberg Notch on the Continental Divide after ascending the 2,000-foot wall above the lake is Heidi Ostrom Duncan. Five miles away and 3,000 feet lower lies Swiftcurrent Lake, the Many Glacier Hotel, and our car. Barely visible in the middle distance is Shangri-La Lake, high on the east flank of Mount Wilbur and Bullhead Point; Shangri-La in the background.

The Iceberg Notch is another classic climb, worked out by Dr. Edwards, who graciously guided the Over-the-Hill Gang on our first time up. There is only one short nasty pitch, but it can be done without support if you are very careful. It is elemental to always be very careful.

Views from Iceberg Notch are just like all the other high places in Glacier, indescribable. Doesn't matter whether you're looking east down the Many Glacier Valley, to the north where Lake Helen lies far below the 4,000-foot cliffs under Ipasha Peak and the 10,000-foot crest of Mount Merritt and Old Sun Glacier, or to the west where you can see many peaks of the Livingston Range.

Other than climbing up from the lake, one can also reach the Notch via the Highline Trail to Ahern Pass which lies about a thousand feet below here to the west. The climb down is through a steep slope of large rocks. Dr. Edwards calls that route "Boulder Dash." Getting home from the Notch can be via the eastern wall or by a couple of other routes involving the Highline Trail. They all require an early start, clean socks, and two lunches. Some might include a change of underwear.

*Right:* Bighorns may be my favorite high country animal. These two were on the south face of Mount Grinnell. Mighty pretty in summer, but a cold, cruel world in the winter. We do not see many rams from the major trails with exception of the Highline around Granite Park, and Swiftcurrent Pass.

*Below:* Wolves have successfully re-populated Glacier Park on their own by migrating south out of Canada. Though I've seen quite a few in the wilds and photographed a wild pack near Jasper, Alberta, Canada, I've not had a chance to get a close-up in Glacier. This fine specimen was taken under what current wildlife photogs call a "controlled environment," which translates into "game farm" or "zoo." Am intrigued by these big predators, and friends and I have donated North Fork cabin use to wolf research biologists for the past dozen years.

*Right:* This billy has what I consider the best "goat digs" in Glacier Park. We're talkin' plush here, great views, protection from the wind, king sized padded bed, imaginative interior decorating, southern exposure, running water front and back, plus neighborhood access to a beargrass super market. He lives at 7100 Akaiyan Avenue, Sperry, Montana.

*Below:* Moss campion and snow cinquefoil often hang out together. They compose beautiful visual duets to cheer the weary hiker.

If there was ever an example of "luminous moss," this is it. Climbing the southeast ridge of Rising Wolf Mountain years ago, we looked a couple thousand feet down into a giant, hidden cirque where nestled the rarely seen Sky Lake. We saw that its outlet stream dropped over tall cliffs in misty cascades to the east and eventually we tried climbing up along those water falls from the Pitamakan Pass Trail. Timing had to be perfect because ice and snow hang in here eleven months of the year. We made it, and even if we hadn't seen other wonderful things, this one patch of moss made the day for me.

*Above:* This is as close as we ever get to a group picture. Taken at Upper Grinnell Lake in July of 1994 after we made the descent from Grinnell Glacier Overlook, the low place on the skyline. The ice chute where we used a rope is to the left of the cliffs we couldn't get down. Hernia and all, I'd do it again tomorrow, maybe the highlight of thirty-five climbs in 1994.

*Facing page:* The Angel's Wing near sunset with Mount Gould at the right, and Grinnell Lake far below. Looking south toward Piegan Pass. Were there flowers and wildlife? You had to be there.

There are two grizzlies in this picture, at the distance most hikers prefer. The female and her cub are looking for some final food before going into hibernation. They are below the Garden Wall above Logan Creek with Mount Oberlin beyond. Minutes before, they had been only fifty feet from us and mamma was not happy. We were lucky they didn't charge, because evidence points to the female as being the bear that seriously mauled a man hiking alone the following spring.

*Above:* O'Neil contemplates what God has wrought from the summit of Clements. Going-to-the-Sun Road, Going-to-the-Sun Mountain, and our cars at theLogan Pass Visitors Center are visible.

*Right:* Ptarmigan are beautiful birds that turn white in the winter. Maybe this is where that North Dakotan got the idea of wearing white boots in the snow so nobody could track him.

*Facing page:* At last, we're on top of the Little Matterhorn, early September 1994. When seen from Going-to-the-Sun Road just east of Avalanche Campground, this thing looks like the Swiss Matterhorn. It is not a major Glacier peak, but the horn is sheer and it is a long ways from nowhere.

On four occasions over the years, I organized a group and hiked the six miles up to Sperry Chalet to stay overnight and attack the Matterhorn from Sperry Glacier on the following day. That is the easiest way to do it, but each time, storms moved in at night and ruined the plan. Kay Luding, who ran the Chalet, got so she hated to see me coming because she knew the weather would go to hell.

After years of scouting, the gang finally decided to climb it the hard way from Snyder Basin, then go out across the Glacier to Comeau Pass and back to Lake McDonald in one day. All the smart ones turned back on the cliffs above upper Snyder Lake, and leisurely spent the day watching "the biggest bull moose they'd ever seen," ate gallons of ripe huckleberries, took a nap, and got back to the cars in time for the usual big dinner and refreshments at West Glacier.

The others made the summit but had to use rope getting down, and got ice water in their undies sliding off a 200-foot ridge on the Glacier. Seventy-six-year-old Elmer Searles hurt himself during the slide and hobbled out the last ten miles on guts. I offered him some of my "extreme emergency medicine," but being a Mormon bishop he had to turn that down. Total trip involved over a mile of vertical climbing and 18 miles of walking. Tragically, in early summer of 1996, a young man hiking alone was killed by slipping off a cliff below the Little Matterhorn above Snyder Lake. On our ascent, we went to the right of where he fell because our scouring indicated even the safer route to the south should be attempted only in late summer.

*This page:* Female sheep are shedding winter hair when young are born. This lamb is very new and has not been introduced to his mother's band yet. She and two other ewes were guarding their newborns in cliffs on the south slopes of Mount Henkel.

After feeding, the six went back to safe places on cliffs. The first two mothers jumped a small but fearfully rushing stream and their lambs followed; however, this little guy didn't think he could jump that far, and made sounds to emphasize how he felt. The mother then found him an easier place to jump from. Like goats, sheep mothers sleep on ledges with the lamb behind them against the wall, protected against falling and from view of predators.

*Above*: Most of the hoary marmots are seen above timberline where they sit on rocks whistling at each other, but sometimes they go down into the trees. These mountain woodchucks spend most of their lives hibernating because it's winter up there eight months of the year. Grizzlies sometimes dig them out of burrows. We've seen holes big enough to bury a mini-van and always wonder if the bear or the marmot won. This shot was taken on Lunch Creek.

*Left*: Layers of sedimentary rock formed millions of years ago are exposed throughout the park by glacial erosion. Geologists have great times in places like this along the Sperry Trail near Akaiyan Lake. Large group went to climb Mount Edwards, and it was a fine day for wildlife and petunias. Only five actually reached the summit. I might have made it, but fell under the influence of professional dawdlers.

*Above:* Sometimes when three old mountain goats want to simultaneously use the same piece of ground there is a standoff. This one is on the Gunsight Trail along the north side of Mount Jackson, involving Billy, Dean, and Bob. View is east down the St. Mary Valley. The confrontation lasted long enough for the photographer to take three pictures and eat half an apple.

*Facing page:* Just above this waterfall on Baring Creek is where we turn east off the trail to climb Goat Mountain. The towering east face of Going-to-the-Sun Mountain dominates the whole area. Daughter Heidi was asked to please wear a different colored shirt the next time we took her picture here among the wild petunias. Some thought yellow would be good. Petunia is a generic term adopted by older people who lose their flower ID book.

Four people, who did not wait for the photographer, can be seen walking an ice ridge on Sperry Glacier. The low place beyond the three silhouettes is Comeau Pass, between the massive Gunsight and Edwards mountains. On my top ten list of Glacier's "most fantastic" trail hikes is the two and a half miles from Comeau to Sperry Chalet.

*Above*: If I were a painter, I'd go bananas ten times a day running into this kind of arrangement of flowers, lichens, and colored rocks. We see hundreds of small exquisite compositions like this each summer, making up their own special ecosystem. South buttress of Mount Pollock.

*Right*: Nothing quite like the call of a loon, echoing across a remote mountain lake, one of the most haunting voices of wilderness.
The loons of Glacier are monitored by biologists and seem to be holding their own.

See why we call that ridge southwest of Reynolds Mountain "The Dragon's Tail"? In the distance are Gunsight Mountain, Sperry Glacier, Edwards Mountain, the Little Matterhorn, and way out there, Lake McDonald.

*Above:* Black bears are common in Glacier and used to be fed by tourists along the roads. Now that is forbidden. Modern bears developing a taste for human foods usually have to be killed. This fat fellow is doing late fall berry feeding at Mount Altyn. On a 1996 autumn hike to Bullhead Lake up this drainage, we saw three griz and four blacks.

*Right:* If there is any prey species lower than the Columbian ground squirrel in the food chain it is snowshoe rabbits, because they live above ground all year. The healthiest hare in the park couldn't qualify for a two-day policy from Lloyds of London. Remains suggest this one fell to a great horned owl.

Bob Zavadil and his daughter, Anne Tallant, picked a nice place for a break. Look at that moss campion. We climbed Mount Pollock this day by what is called the "Great Cleft" up through the south cliffs. The cleft is quite easy to ascend and descend because it runs at an angle like a built-in stairway; however, two of the fellows complained God should have made it a little wider. There was a fierce storm that raged for half an hour and we waited it out under a small cliff. Saw bighorn rams, two goats, and a marmot.

*Right:* Elk in Glacier are shier than in Yellowstone. Don't know why. We see few up close. This cow probably had a calf hidden nearby, but did not seem alarmed by our presence. She spends her summers at Preston Park.

*Below:* The Over-the-Hill Gang isn't alone in loving that beautiful remote place we call "Eden." Almost every kind of animal found in Glacier probably visits here because of the diverse location and attractions. On this day we saw snowy goats along the northwest rim. Then, while eating lunch by a waterfall, located a band of bighorn rams. I caught at least five of them with this 300mm telephoto shot. Earlier that morning I blew my first and only chance for a wolverine close-up. He was on a rock in the sun, snarling in typical carcajou fashion. Five fast shots were blanked by incorrect film loading. No matter what the others say, I did not bawl like a baby.

*Above:* Almost every book and magazine about Glacier has "the goat above the lake shot," usually Ellen Wilson Lake, or Hidden Lake. This Ellen Wilson shot gives a little different angle because the billy is not standing in Gunsight Pass, but up on the south ridge of Gunsight Mountain.

*Facing page:* If there was a road to this place, it would be a featured attraction in every first class publication from *Montana Magazine* to *National Geographic*. The half-mile gorge on Nyack Creek is unnamed but it is a dandy. Getting there involves wading the Middle Fork of the Flathead up to the trap-door on your long johns, falling in the creek and losing your ice axe, then listening to friends bellyache for five miles because they had to help fish it out with a long stick. Great place.

Colors, colors everywhere in Glacier Park. If it isn't lichens, flowers, moss, or shrubs, it is rocks. This display is typical of the multitude of geological formations viewed constantly while hiking. Happily, the rock wall's beauty is accented here by vegetation, and just about anyone can reach this spot on Grinnell Glacier Trail. Sightseeing boats on Swiftcurrent and Josephine lakes get you close or you can hike from Many Glacier Hotel.

*Right:* The Lewis monkeyflower was named in honor of Captain Meriwether Lewis who led the 1804-06 expedition through Montana, trying to find out exactly what President Jefferson had bought at Napoleon's going-out-of-business sale. Lewis visited the high plains east of Glacier but did not venture into the mountains. Lewis monkeyflowers come in two colors, red and yellow. These along Lunch Creek.

*Below:* So many of the plants in Glacier have life-links to others, from the lowest lichens, to mosses, up to orchids, and giant cedar trees. Some of those relationships are mutually beneficial, some are not. Noticed delicate little plants living off this elephant shaped moss, 7,000 feet up on Cataract Creek. Glacier's plants run the complete list from rain-forest-type ferns and orchids in the west to the hardiest arctic lichens.

*Facing page:* On top of Reynolds Mountain looking north to Clements Mountain with Mount Cannon at the left, and the south end of the mighty Livingston Range in the far distance. Hidden Lake is 2,700 feet below. Reynolds is on the Continental Divide so its waters flow to the Arctic Ocean via Hudson Bay, while westward, at Bob Zavadil's back, to the Pacific.

From Triple Divide Peak, eleven miles southeast of this point, waters flow to three oceans: the Arctic, the Atlantic, and the Pacific. There are unconfirmed stories of some guys drinking a lot of water and climbing up there so they could stand on the top and go potty in three oceans at the same time. It was very windy the day the gang was there.

Reynolds is one of the best known peaks in the park because it sits there dominating Logan Pass south of the Visitor Center. A couple million people see it every year. My first ascent was thirty years ago.

Glacier's mountains are named after early-day explorers, well-known Indians, historic events, famous white people, and a few for minor Washington, D. C. politicians. Names like Reynolds, Gould, Cannon, and Clements are not Indian, but we have Heavy Runner, Piegan, Going-to-the-Sun, Little Chief, and dozens of others. The Over-the-Hill Gang has climbed peaks like Bad Marriage, Never Laughs, and Almost a Dog. My favorites use Blackfeet and Kootenai Indian phonetic spelling so we get Kupunkamint, Akaiyan, Otokomi, Sinopah, Natahki, Pitamakan, and Akokala. There is a fascinating book on Glacier Park names, *Place Names of Glacier/Waterton National Parks* by Jack Holterman.

If anyone has actually counted all the named and unnamed peaks in the park, they've neglected to tell us. Individually or together, the TOHG has climbed over a hundred, some several times. Over-the-Hill Gang champion, Ivan O'Neil, is well beyond the 100 figure.

*This page:* Saw fourteen bighorn rams on Swiftcurrent Ridge below the lookout in August. Rams fight over dominance or for practice any time of the year, but the mean battles are in late fall for love. These young males were hitting heads hard. Maybe a personality thing. Prior to slamming their well-insulated skulls together, rams rear up on back legs for more power. Afterwards they stagger around like a three-martini luncher.

*Left:* It's a seventeen-mile trail hike around the back side of the lovely St. Mary Lake, but the Reverend George Tolman carried his fishing rod "at the ready" every step of the way. This area shows amazing moss formations; but I suppose when a guy carries a fishin' pole that far, he loses interest in moss.

*Below:* Momma moose is looking things over. Ears up or forward, "OK." Ears laid back, "You should find a tree."

A daring human conquers the last horrendous pitch on the dreaded south face of Altyn. After making those last few feet, he was rewarded by the sweeping panoramas of Swiftcurrent Lake, Mount Wynn, Mount Allen, and the 10,000-foot summit of Siyeh on the skyline—and that's just the south view. When asked what he was doing, the jubilant fellow said he was celebrating his sixty-fifth birthday, and heard this spot offered the best possible views of the Many Glacier Hotel roof. *(Photo of her father by Heidi Ostrom Duncan)*

*Above:* There were four coyote pups here but two minded their mother and ran across Canyon Creek with her. Son Clark and I called the group in by doing the "dying rabbit screech" with a blade of grass. Mamma coyote quickly determined the guy in the silly hat was too big for a rabbit, but these two hung around a minute. This was on Cracker Flats, two miles east of the Many Glacier Hotel. The dying rabbit screech is now forbidden in national parks, but it wasn't then.

*Right:* Dale Haarr and Ed Mall were among a large group who believed me when I said, "I'll bet it would be easy to climb down from Grinnell Overlook to the glacier, then hike out to the Swiftcurrent Trailhead." The good news is that this was a fantastic descent as far as scenery and wildlife were concerned. Look! In perfect view is the Angel's Wing, Upper Grinnell Lake below Salamander Glacier, Mount Gould on the Continental Divide, and much much more. The bad news is we could not find a way through the cliffs right above the lake. That's when I got a hernia helping Charlie Bleck down an ice chute with the rope.

Had to hike miles through this kind of October scenery in 1994 while the gang wouldn't stop for me to set up a picture because "...we've got a long way to go." On the Highline Trail, halfway between Ahern Pass and Granite Park Chalet, I finally stopped anyway. Forest fires were burning in the north of the park and the sunset was magnificent. Took some more shots and got left behind. There is no way those old fogies are going to get every peak in the park climbed, but the problem is they haven't figured that out yet.

In November of 1995, the gang hiked up Going-to-the-Sun Road to Logan Pass after winter snows had closed the road to motor traffic. On the way up we saw fresh black bear tracks near the Triple Arches. Coming down, Elmer Searle was leading when he yelled, "George, get down here with your camera. There's a black bear behind those rocks." Dean Dahlgren was with me as we hurried past the others and started sneaking and peering over the side. Suddenly we were face to face with a black bear all right, a black grizzly bear and her cub. Bad news. She partially stood up and expelled air from her lungs with such velocity it made a high-pitched whistle through her teeth. This was followed by a very deep and almost soft growl as she went back on all fours. Besides the camera, I was carrying a climber's axe and binoculars. Began passing everything to Dean so I could get my bear spray, and still keep eyes glued to the bears. He said, "George, I've only got one arm." And he does, too.

We talked softly and backed up until she relaxed and started down the steep sidehill. My major thought was, Ostrom, how dumb can you get?

Did a couple shots up close, but didn't have time for proper settings and they turned out overexposed and, of course, blurry.

*Above:* Some of the gang led by Dr. Edwards are descending his secret goat trail, which runs up over a high rocky ridge northeast from Ptarmigan Tunnel, and in this spot goes down to Red Gap Pass. So many animals use this route it looks like a mountain goat I-90. These ranges on the eastern front are more barren in appearance than those along the Divide, but they are a floral paradise with abundant wildlife, many springs, lakes, and creeks. That far peak looks a little reddish, but is called Yellow Mountain. Maybe it's the light.

*Facing page:* Sometimes Glacier Park can seem excessive with its dazzle, like a beautiful woman with too many jewels. On this morning we hiked eight miles. First along one of the prettiest alpine lakes in the world, through forests, across lush green meadows with yellow glacier lilies, past the thundering Rockwell Falls, along the gorge, then up through the red-rocked gardens below Cobalt Lake. It was July and the view is west down Park Creek on the Pacific side.

On the Piegan Pass trail. Awesome walls of Mount Gould form background. One snowbank looked like George Washington with his cape. After descending below treeline above Grinnell Lake, the huckleberries were so thick, Heidi picked enough for a pie without stopping. Caught the last Josephine Lake boat, and I sat on the berries while getting into the car.

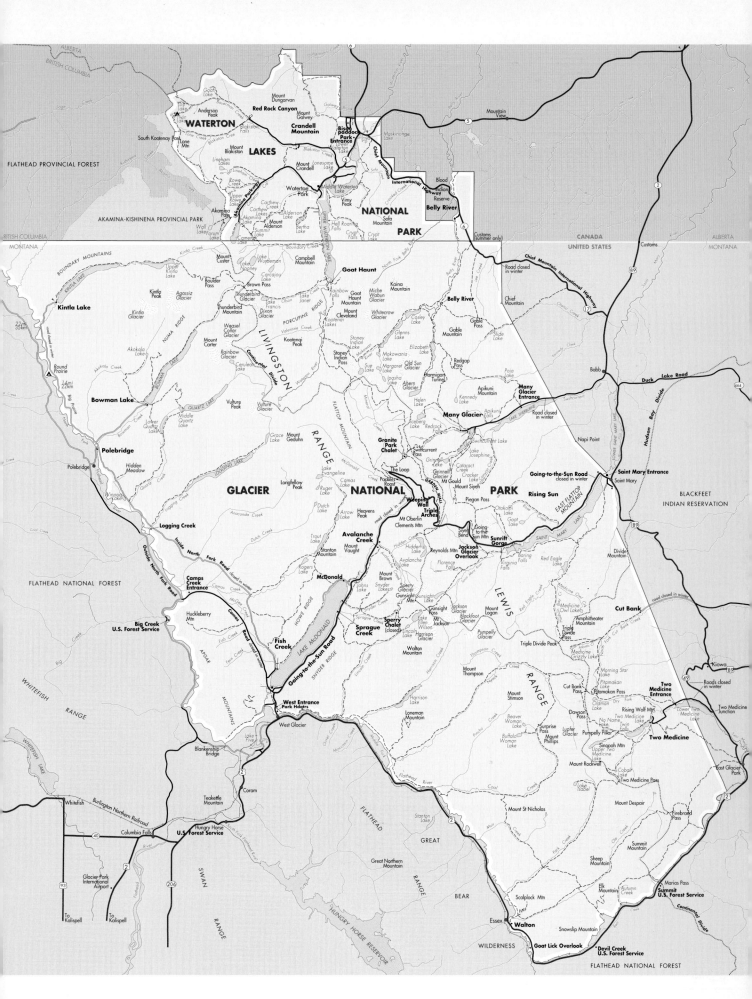

## NOTES ON GEAR AND PACKS

Proper clothing is important because it can snow on any day of the year in Glacier Park's high country. Rain with a wind can quickly produce hypothermia. Number-one item is footwear because feet have to be comfortable, and there must be good traction. For off-trail travel, we like good ankle support and ankle protection because of fractured rock. Most over-the-hillers invest in top name-brand hiking boots, but these are not needed for most trails. For outerwear, flexibility is the key word. Warm light jacket's a must. Might start in the morning wearing long pants, go to full wraps, and finish in shorts. Depends on weather and weather is not dependable. Hats are nice, at least a warm knit one in the pack.

A day pack contains sun screen, Scout or Swiss Army-type jackknife, insect repellent, dry socks, flashlight, lunch, extra hiker's rations, first aid kit, windbreaker and/or raincoat, plus water bottle, fire starter, compass, map, and *gloves*. Gloves are for warmth, but equally important, to protect hands against sharp rocks, mainly on descents. A few of us tote those lightweight space blankets.

Over-the-hillers all have a cannister of Counter Assault bear spray within easy, immediate reach, but we've never actually had to use it. Optional equipment carried by some members includes ice axe or walking stick. Water shoes are handy if there are stream crossings. Binoculars are good, especially the light ones. Usually have a couple of ropes for emergencies. I wouldn't be caught dead without a camera. Air is thinner up high so sun can be a problem on eyes. That calls for "shades" with ultra-violet filtering.

One member always has a small plastic bottle of "snake bite medicine," knowing full well there are no poisonous snakes in Glacier.

## THURSDAY OVER-THE-HILL GANG

The Thursday Over-the-Hill Gang has become a part of Glacier National Park. Each spring the gray-haired members follow receding snow up the mountains just as surely as the migrating elk, and they keep going up until fall storms drive them down. Wilderness is an integral part of their lives. Waiting for the next spring means cross-country and downhill skiing. They come from all walks of life; there being no set standards, the membership includes several lawyers. Half a dozen are former military pilots.

Media publicity and word-of-mouth has brought hikers from different countries as well as many states, wanting a trip with the Over-the-Hill Gang.

Lately there are often three to five separate hikes or climbs at once. We have the "B" Team, those most dedicated to getting on the top and are in better shape. Other than three oldies, it includes "the kids," those under 60.

The "A" Team may or may not go to the top, might stop by a waterfall for a nap, or stop to fish. They take more pictures, have a deeper aesthetic sense, and seem to dawdle. Sometimes the team members spend all day together. Other times they split.

"C" Team are those getting over a bad hind leg or not feeling up to par. All teams contain interchangeable parts. The C Team may hike only a few miles, or drive cars to pick up the others, but they go.

For years the original group was Ambrose Measure, Spencer Ryder, Hi Gibson, Harry Isch, and Ivan O'Neil. Measure began hiking in the 1920s and was still at it seventy years later. Thursday became *the* day in 1977. By the early 1980s, Ostrom and Walter Bahr showed up.

In the late 1980s, breakfast had ten regulars. Size increased after 1986 when Elmer Searle, Pat Gyrion, and Bob Dundas joined. In the 1990s, the sex barrier was broken by two ladies, Bobby Gilmore and Sandy Everts. Got more males, too: Dean Dahlgren, Hank Good, Jack

Fletcher, Gene Jacobsen, Dean Jellison, Jim Eagen, Bob Zavadil, Roger Dakin, Vern Ingraham, Dee Strickler, Roger Somerville, Jim Galvin, Ted Rugland, Dale Haarr, Bill Harris, Charles Kempner, Charles Bleck, Bob Kroger, Randy Heim, Ed Mall, and Larry Sandefur. Visiting minister George Tolman has done many climbs, as have Ostrom's daughter Heidi Duncan, and son Shannon. Editor Brian Kennedy of the *Hungry Horse News* goes and so does County Attorney Tom Esch. Members have been accompanied by sons, daughters, and in-laws. Ivan's daughter Cheri Matthews and her husband, Bob, show up every year from California. Bahr and Galvin kids, as well as Butch Jensen and offspring join now and then.

About half the Gang are retired. There are no requirements except the ability to "be a good team player," never wander off by yourself, or leave another member alone off-trail. Our wives each understand that when her husband goes to "the big Alpine meadow in the sky," the services are not to be held on Thursday.

## BOOKS AND GROUPS FOR NATURE LOVERS, HIKERS, AND CLIMBERS

*A CLIMBER'S GUIDE TO GLACIER NATIONAL PARK* by J. Gordon Edwards, 1995. The 352-page climbing bible from the patron saint of Glacier Park mountaineers. Besides detailed description of routes on the peaks, Dr. Edwards covers wilderness dangers, Park rules, protecting the environment, history of climbing in Glacier, and an extensive list of selected literature. Contains diagrams, black and white photos, and eight pages of color.

*GLACIER COUNTRY: MONTANA'S GLACIER NATIONAL PARK* by Bert Gildart and others, 1990. This is the best all-around source for info on the park's geology, human history, micro-climates, and total environment. Published by American & World Geographic Publishing in Helena. Has fine photos and excellent text. There are two other books from this Montana Geographic Series which have outstanding photos and info related to Glacier Park. They are *Montana Wildlife*, and *Montana Mountain Ranges*.

*THE TRAIL GUIDE TO GLACIER NATIONAL PARK* by Erik Molvar,1994. Maps and lists trails, distances, altitude changes, campgrounds, and safety tips.

*PLANTS OF WATERTON-GLACIER NATIONAL PARKS* by Richard Shaw and Danny On, 1979. There are over 1,000 species of just vascular plants in Glacier so no book lists them all, let alone the lower species of mosses and lichens. This one covers 220 plants, representing twenty-two of the "families." There is also an excellent book, *Alpine Wildflowers,* by Dr. Dee Strickler, who has climbed on occasion with the TOHG.

*PLACE NAMES OF GLACIER/WATERTON NATIONAL PARKS*, by Jack Holterman, 1985. A totally fascinating piece of research. Holterman knows the last 200 years of history surrounding Glacier Park and the adjacent Indian tribes in detail like no one else. This is not just a priceless information source, it is a captivating read.

*ALONG THE TRAIL: A PHOTOGRAPHIC ESSAY* by Danny On, 1979 and reissued; text by David Sumner. Danny On was an internationally known forest scientist, war hero and friend of all who knew him, killed in a skiing accident in 1979. His family gave the Glacier Natural History Association the rights to publish a book of his photos.

*HIGH LIFE: ANIMALS OF THE ALPINE WORLD* by John Winnie, Jr., 1996. John Winnie is not just a top photographer, he is also a studious zoologist, willing to suffer arduous climbs and even frostbite if that's what it takes to get the picture. His book is the best and most interesting coverage on the North American alpine animals, the predators that stalk them, and how they all miraculously survive in that beautiful but tenuous world of rock and ice.

The GLACIER NATURAL HISTORY ASSOCIATION is a cooperating non-profit corporation, associated with the National Park Service. It operates bookstores in the visitor centers and maintains a headquarters at West Glacier. The profits of the Association go to further the Interpretive Division in Glacier Park. Many maps, books, and pamphlets are available. The address is Box 428, West Glacier, MT 59936. Phone (406) 888-5756.

THE GLACIER MOUNTAINEERING SOCIETY is a formally organized group that does all types of climbing in Glacier Park and elsewhere in North America. They are into technical ascents and hold an annual "GMS Week" in July that attracts Glacier alpinists from across the nation. Recognition awards for climbing accomplishments are made at the GMS and Old-Timers Banquet at St. Mary. Some TOHG members belong, and one who works cheap emcees the banquet. The address for GMS is P.O. Box 291, Whitefish, Montana, 59937, c/o Dennis Twohig.

# INDEX

## ABOUT THE AUTHOR

**G. George Ostrom** is a 69-year-old journalist who has spent his life near Glacier National Park. He worked five summers as a parachute firefighter, or smokejumper, for the Forest Service while at the University of Montana, and spent two years in Washington, D.C. as a Kennedy "New Frontiersman" helping write the Wilderness Bill. His photographs and stories have appeared in national magazines including *Sports Afield*, *Field and Stream*, *Saga*, *National Parks*, *Camper*, *Sailing*, *Bugle*, and *Sports Illustrated*.

George has received numerous honors for newspaper writing and radio broadcasting, and is the only reporter to cover the eight fatal grizzly attacks on humans in Glacier Park since 1967. His on-scene photos were used in the book, *Night of the Grizzlies*. Perhaps he is best known for a weekly column, "The Trailwatcher," started in the *Hungry Horse News* in 1963. Reprints have appeared in major newspapers from the *New York Times* to the Seattle *Post-Intelligencer*. Over the years, his column has won top state and national awards, and in 1996 it was selected by the National Newspaper Association as best weekly humor column in the United States.

He got into the camera habit hiking with the late photographers Hal Kanzler and Danny On, a former smokejumper buddy. George said, "As long as I was helping carry all their darn gear, figured I might as well try a few shots." He has since given slide shows to thousands on the subjects of wildlife, hiking, and climbing. After the deaths of Kanzler and On, George hooked up with the "Thursday Over-the-Hill Gang," a "well-disorganized local group of overly mature people" who average thirty-four days a year on the peaks of Glacier Park.